MAN AS DNA

MAN AS DNA

Bridges to the New Reality

by

Earl Hubbard

Philosophical Library
New York

Library of Congress Cataloging in Publication Data

Hubbard, Earl.
Man as DNA.

1. Man--Addresses, essays, lectures.
2. Reality--Addresses, essays, lectures.
I. Title.
BD450.H812 1982 113 82-11255
ISBN 0-8022-2408-3

Published 1982 by Philosophical Library, Inc. 200 West 57 Street, New York, N.Y. 10019.Manufactured in the United States of America.

Overseas distributor: George Prior Ltd. 52-54 High Holborn, London WC1V 6RL England.

Dedication

To Jacqueline Hériteau,
for whom this book was written.

TABLE OF CONTENTS

Introduction

bridge la: a structure carrying a pathway or roadway over a depression or obstacle b: a time, place, or means of connection or transition
—Webster's New Collegiate Dictionary

To spend a lifetime as an artist—as I have done—is to be a pilgrim, a seeker of meaning. The backbone of all culture is formed by the evolution of religions, for they represent the broad mainstream of human thought. We can even say that the evolution of humankind is but the primary level of a cosmic consciousness of our identity as an insatiable hunger for—call it meaning, or purpose, direction. It is the constant yearning from which all that we know and are is spun.

So, we can say that to be an artist is to bear witness to the search for meaning: this is what art is all about.

To be an artist in this century has been to see the utter dissolution of the image of man as purposeful. Our age has been the black hole into which the Renaissance has been swallowed. To see Michelangelo's image of David atomized to the splatterings of Jackson Pollock; to see the walls of museums and galleries hung with spiritless epitaphs for the corpse of a past, twice buried in world wars; to hear the ghoulish cheers and chants of a charlatan band of critics claiming what was in fact a funeral to be a celebration of life; to see the undertakers of art auctioning off to the highest bidders the sleezy graffiti of a "popped" culture—has been a chilling experience. To look for sustenance in the galleries and museums of contemporary art is to discover a desert. Since World War Two art has been a sterile world celebrating a vacuum it ardently articulated.

The subject matter of art is reality. Art has accepted this age as the end of a past concept of reality.

Our concepts of reality are constantly changing. Different ages in history represent different concepts of reality. Traditionally this subject matter has been defined by religion. In search of a new concept of reality I turned to the church. I discovered largely empty buildings echoing what through endless repetition have become equally empty cliches. Or the churches were silent, their clergy out on the streets protesting the iniquities of various minorities or the assumed moral bankruptcy of a nation attempting to be the conscience of the world. The churches have become monuments to a reality no longer acceptable as valid or relevant.

Still, as silent sanctuaries in which a fragmented population could retire to escape too much that was happening too fast, they have served as reminders of a time when we thought we knew who we were and what the world was all about. They have reminded humankind of the constant challenge each of us is compelled to meet—that of seeking to learn who or what we are. Churches have ceased to be symbols of affirmation.

They stand in a new age as great and small question marks.

With the understanding that the church now represents an old reality I set out to see the new reality. I heard two prophetic messages. The first was that of Werner Heisenberg. Winner of the 1932 Nobel Prize in Physics for his Uncertainty Principle, Heisenberg said in his autobiography, "Atoms are not things (but processes). The electrons which form an atom's shell are no longer things in the sense of classical physics, things which could be unambiguousuly described by concepts like location, velocity, energy, size. When we get down to the atomic level, the objective world in space and time no longer exists and the mathematical symbols of theoretical physics refer merely to possibilities, not to facts."

And he said, "The very attempt to conjure up a picture (of elementary particles) and think of them in visual terms is wholly to misinterpret them."

"The atom is not a thing but a process," is such a simple statement, as simple as the first four notes of Beethoven's Fifth Symphony. As it echoed through the succeeding years I came to realize the nature of the new reality Heisenberg's statement heralds is metaphysical. For if the atom is not a thing composed of things, but a process, that means everything is a process composed of processes.

That means the cosmos is procedural as opposed to being mechanical. Sir James Hopwood Jeans, the distinguished English mathematician, physicist and astronomer, put it this way in his Rede Lectures:

"Today there is a wide measure of agreement, which on the physical side of science approaches unanimity, that the stream of knowledge is heading towards a nonmechanical reality; the universe begins to look more like a great thought than like a great machine."

Sir Arthur Stanley Eddington, the distinguished

British astronomer and physicist, put it in a comparable way: "The stuff of the world is mind- stuff."

The second prophecy announcing the new reality came from Francis Harry Compton Crick, James Dewey Watson and Maurice Hugh Fredrick Wilkens. These three shared the 1962 Nobel Prize in Medicine and Physiology for their unraveling of the structure and function of deoxyribonucleic acid (DNA). DNA we now know is the key substance in the transmission of hereditary characteristics from generation to generation. What the DNA discoverers said, in effect, is all life evolves according to a plan. The international community of scientists' acceptance of the validity of their discovery of a plan lays to rest the age-old argument about whether life occurs by accident or by plan. But the implications of the discovery of a plan extends far beyond the settling of an old argument. A crucial question arises: If all life evolves according to a plan, where does that plan come from? The rational answer is that all plans evolve from previous plans, which evolve from previous plans. The discovery of a plan in the DNA molecules governing life opens up the very real possibility that there is no such thing as randomness, that **all** is planned.

For someone raised on the mythology of self-made-man in a mechanical universe this discovery came as a great shock. If I am not responsible for my identity what am I responsible for?

Though the general public may not spend much time considering the implications of the DNA plan, by a remarkable coincidence the questions our age faces with anguish relate to identity and personal responsibility. The English-American poet W.H. Auden called our age the Age of Anxiety. The nature of our anxiety is our identity crisis. The question "who am I?" is the question "what is reality?" We have lost our past sense of reality, of who we are, and "all the king's horses and all the king's men can't put it back together again." But given Heisenberg's observation that the atom is not a thing, it is apparent all that has been lost is a conception of reality—for there is no reality of "things" to lose.

With the awareness given us by Crick, Watson and Wilkens that all life evolves from a plan—which invites the supposition that all plans are composed of plans, composed of plans, etc. etc. etc.—and therefore there is no randomness, the void of our despair can be filled with the hope we are part of a purpose—albeit an unknowable purpose—for presumably it is ever-evolving, and thus never definable.

The following essays log my odyssey into the new reality proposed by the discoveries of Heisenberg, Crick, Watson and Wilkins. I have no doubt the essays are as accurate as the early maps of the ancient mariners on which were written such ominous warnings as "here there be dragons." Indeed these are not maps to be followed so much as they are bridges. Once crossed, the reader is on his or her own, to explore what presently appears to be an almost infinite frontier.

Earl Hubbard

New York, N.Y.

Bridge 1

INFORMATION

Physicists have discovered that the material world is nonmaterial. It is as though in seeking the ultimate immutable particle from which it was assumed the material world is built physicists broke through a wall they didn't know was a wall and discovered nothing on the other side. The word "nothing"—NO THING—signifies they didn't find what they had been looking for, an immutable particle. They did find a concept. They found that the idea of a material world is a wall between two different ways of thinking about reality. On the old side of the wall of materiality it was assumed that "reality" was outside what we thought—that a tree falls whether we are there to observe or not. On the new side of the wall of materiality it was discovered that reality is exclusively what we think. There is a profound difference between accepting reality as existing, present, whether we think it's there or not—and accepting reality as not existing outside our thought, but as thought itself. Until this discovery was made we thought we *saw* reality, *felt* reality and *heard* reality. But according to physicists, we *think* reality. What we experience, they say, is our thinking—it is the only substance and all that is real.

Put the other way around, the world around us *is* our thought.

According to Webster's New Collegiate Dictionary to think is "to form or have in mind."

We cannot form or have in mind, nothing. We cannot think nothing. The something we form or have in our minds, the stuff of our thoughts, we call information. We think about information. To have information we have to be informed.

According to Webster's New Collegiate Dictionary to inform is: "to give material form to 2a: to give character or essence to (principle which informs modern teaching) b: to be the characteristic quality of: ANIMATE (the compassion that informs his work) 3: GUIDE, DIRECT."

To be given information then is to be given form, character, essence, characteristic quality (personality), to be animated, guided and directed.

In Webster's Collegiate Thesaurus, the synonyms for "inform" are: "1. FIRE 2. animate, exalt, inspire, rel. (relative terms) imbue, infuse, leaven, permeate: enlighten, illuminate; endow, endue."

If reality is our thinking, and to think we must be informed, it would seem that the act of creation is an act of informing. The New Testament's way of saying it is, "In the beginning was the Word." (John 1:1)

The life that fires, animates, exalts and inspires us is information. It is information that imbues, infuses, leavens, permeates, enlightens, illuminates, endows and endues us. To be born is to be informed. To live is to be informed.

If life is information, our bodies must be the processors of information.

Webster's New Collegiate Dictionary defines processor as "1: one that processes (food processor) 2a(1): COMPUTER (2): the part of a computer system that operates on data—called also central processing unit b: a computer program (as a compiler) that puts another program into a form acceptable to the computer."

If life is information, our bodies' needs must be informational. Food, air and water must be forms of information. Our digestive system and lungs must then be processors putting the information of food, air and water into forms acceptable to our computer-like bodies. The sensual needs of our eyes, ears and tongues must also be informational. Eyes, ears and tongues must be processing the information they receive into forms acceptable to our computer-like brains.

Without sunlight there would be no life on earth. Sunlight must then be our primary source of information. Darkness to us means the absence of information. When we are confused we often say, "We are in the dark."

Coal, oil and natural gas, the energy resources on which our industrial-electronic age depends are forms of stored sunlight. Coal, oil and natural gas must then be forms of information essential for feeding the informational needs of our industrial- electronic systems.

What we call energy is apparently information.

If life is information it would appear that what we call our industrial-electronic systems have life. The sunlight that informs our biological bodies informs their non-biological bodies.

Sunlight comes from the sun. Scientists surmise that the sun is composed of hydrogen atoms that in a high state of activity are converting into helium atoms. The sunlight we feel and see, they say, is the energy released by this process of conversion.

As far as astronomers can see into the universe, there are luminous hydrogen clouds billions of miles long and billions of miles wide. Some of these clouds appear to be condensing into stars. They therefore assume the first star condensed out of a hydrogen cloud billions of miles long and billions of miles wide, billions of years ago.

Solar systems, astronomers conjecture, also condense out of hydrogen clouds. However, these are clouds that have been fertilized by supernovas. A supernova is an exploding star. It is thought that after all the hydrogen atoms in the star have converted into helium atoms but before the star explodes, it collapses in on itself creating a furnace. In this furnace are forged the elements essential to the synthesis of the minerals that compose planets. When the star explodes it scatters these elements into a hydrogen cloud, in effect inseminating it with the capacity to give birth to solar systems. Another way to say this is the hydrogen cloud is informed of the possibility of conceiving solar systems.

Our earth then is informed—conceived—as our bodies are. If information is life, our earth and all planets are alive as our biological bodies are.

Scientists conjecture the elements that compose our bodies as well as those of the birds and beasts, reptiles, fish and vegetation of this earth along with the planet itself were conceived in the same furnace of the same collapsing star. It would appear that we, the earth, and all that is on it, are composed of the same stardust. It would also appear that the substance of what we call the universe is what we call hydrogen. Astronomers claim to see and have photographed what they call hydrogen clouds but no one has ever seen a hydrogen atom. No one has ever seen an atom of any kind.

This is the "nothing" scientists discovered when

they broke through the wall of materiality. Werner Heisenberg won the 1932 Nobel Prize for discovering that on the atomic and subatomic levels "no-thing" is there. He said in his autobiography, "Atoms are not things." Later in that same autobiography he referred to atoms as "processes."

If information is life, the atomic process must be the process of informing. The way this informational process works is indicated by a discovery made by Francis Harry Compton Crick, James Dewey Watson and Maurice Hugh Fredrick Wilkens. These joint winners of the 1962 Nobel Prize in Medicine and Physiology discovered that all that we call "life" is built according to a plan called DNA (deoxyribonucleic acid). DNA, they say, programs cells to do what must be done to build what we call "living bodies." Their discovery indicates DNA informs the cells through the process of programming.

Something must inform DNA. Something must program DNA. Ultimately that something must be what we call (but have never seen) the hydrogen atom.

Werner Heisenberg said, "Atoms are not things but processes." It would appear that that process is programming. It would appear that the process of informing the cosmos is the process of programming.

What we mean by "atoms," then, are programs programmed to program the evolution of the aggregate or groupings of atoms we call molecules. Molecules then must be both programs and programmed to evolve either non-biological or biological forms of life. Which in turn are programmed to program our reality.

We know our reality is ours alone. Each of us has experienced that awful sense of utter aloneness, that inability to become part of someone else's reality or to draw someone else into ours. Such experiences are at least frustrating and at most terrifying. Currently we call these experiences "break-downs in communication."

It is as though each of us is a unique program designed to process programs in a unique fashion.

The applicability of this awareness to personal relationships, not to speak of international relationships, is apparent. What we have been calling breakdowns-in-communications between individual people and nations may now be understood as failure to allow for differences in programming.

According to Webster's New Collegiate Dictionary the word "program" is defined as: "1" a public notice 2a: a brief usually printed outline of the order to be followed, of the feature or features to be presented, and the persons participating (as in a public exercise, performance, or entertainment) b: the performance of a program; esp. a performance broadcast on radio or television 3: a plan or system under which action may be taken toward a goal 4: CURRICULUM."

In the context of our acceptance of ourselves as specific programs Webster's definition, "a public notice," could be interpreted as a public awareness of ourselves as a specific reality or what we commonly call an individual. The words "order," "exercise," "performance," "plan," "system" or "curriculum" all indicate a specific sequence. To accept our reality as a specific program is to accept our reality as a specific sequence.

In accepting our reality as unique we are identifying ourselves as part of a sequence. To say something is different is to say it's part of a sequence. Otherwise it's not different but the same thing. How else can we account for our diversity except in terms of sequence? This is how reality comes to us, in a sequential fashion. This is how we perceive what we call evolution.

Programs are sequential.

Apparently we are programmatic parts of the program we call evolution.

It appears that the wall physicists broke through is a wall between two ways of thinking about ourselves. On the old side of the wall of materiality we thought reality—a material world—existed whether we existed or not. On the new side of materiality we discovered reality does not exist unless we exist. On the old side of the wall of materiality we thought of ourselves as unprogrammed. On the new side of the wall of materiality we see ourselves as programmed.

The differences between these realities on the opposite sides of the wall of materiality are profound.

Bridge 2

FREE CHOICE

When we talk about the future we are attempting to answer the question "What will become of me?" This question has within it the question "Who am I?" For what we are to become has its roots in what we already are. For example, when we futurize over what will become of an acorn, once it's planted, we say it will grow into an oak tree. In saying this we are assuming that the acorn represents a particular potential—that of becoming an oak tree. This assumption represents an attitude towards the present. It shows that we conceive of the present in terms of potentiality. We think of the acorn not as an oak tree but as having the potential of becoming or growing into an oak tree. This is exactly how we think of ourselves. We think of ourselves as potentialities. Our sense of identity is largely our sense of our potentiality.

The difference between our attitude towards the potential of an acorn and the potential of ourselves is we imagine we know what an acorn will become while we imagine we don't know what will become of us. However, our attitude towards history is comparable to the one we have towards the acorn. When we read about the young "acorn" Beethoven we know he will grow up to become a "great oak" composer. Part of the fun of reading biographies is that we know how the story ends. We are amused to read about how the hero agonized in self-doubt. From the perspective of history all anxiety appears to be futile self-abuse. It appears to be as ridiculous for Beethoven to have agonized over becoming a great composer as it would be for an acorn to agonize over its capacity to become an oak tree. For all we know, acorns may very well agonize over their capacity to become oak trees.

That isn't to say that Beethoven's agony wasn't a factor in the process of the composition of his music anymore than the possible agony of the acorn may be a factor in the process of its evolution into the specific kind of oak tree it eventually becomes. Agony, however, if it is a factor in the evolution of a Beethoven and/or an oak tree, is the only indication of free choice that is apparent. From the perspective of hindsight, everything that happens appears to have happened inevitably.

In general we accept the evolution of nature and our own historic past to have been inevitable. At the same time we assume our own personal evolution is not inevitable. Cosmologists, for example, assume that everything that happened, happened in an orderly sequence.

Cosmologists don't say things like, "Boy, it sure was a lucky thing that the cosmos evolved."

Nor do meterologists say, "Thank God the sun rose this morning."

No, nature is accepted as an orderly sequence of inevitable events, whether we know what the exact sequence is or not. We assume all that we call "nature" is the evolving realization of some predetermined plan.

According to Harry Compton Crick, James Dewey Watson and Maurice Hugh Fredrick Wilkens, who shared the 1962 Nobel Prize in Medicine and Physiology for their unraveling of the structure and function of deoxyribonucleic acid (DNA), all life forms are built according to a plan. Plan, they say, is what DNA does. It tells the cells how to build a body, whether it be the body of an oak tree or the body of a Beethoven. According to this information, all life forms would appear to be predetermined. A plan indicates predetermination.

Where did the plans for DNA come from? If all life evolves according to a plan isn't it reasonable to assume all non-life evolves according to a plan also? The cosmos itself must evolve according to a plan.

When we think of ourselves most of us most of the time assume that we do what we do intentionally, deliberately having made a conscious choice. At the same time, most of us assume that oak trees, birds and galaxies do what they do because they have to do it, not because they intend or choose to do it. Most of us assume that we do what we do out of choice, but that nature fulfills a predetermined plan. But the question that inevitably arises is how can we who are part of the cosmos and a very small part at that, believe that we intend or choose whatever we do if the rest of the universe is pretermined?

To answer that question let's go back to the acorn. The acorn, once planted, becomes an oak tree just as we expected it would. In fact every time we plant acorns we expect oaks to grow and either an oak tree does grow or nothing grows. We have not planted an acorn and had a tulip, or an elephant, or a whale, nor even a person, grow up out of an acorn. Therefore we assume that to plant an acorn is to predetermine an oak. Or almost predetermine an oak. Something could go wrong with the acorn—a squirrel could dig it up, or a wild boar might eat it or various kinds of insects attack it or whatever. But what does this mean? Was this particular acorn predetermined to feed squirrels or wild boar or insects or whatever, or simply to rot and enrich the soil? And are some squirrels, wild boar, insects and whatnot including the bacteria involved in the rotting process of an acorn, predetermined to eat certain acorns?

Or—say the acorn does grow into a great oak. It might be cut down and turned into lumber to build a house, or paneling for a library, or it might be cut down and sawed into firewood to burn in stoves for heating or for cooking. It's also possible that it won't be cut down at all but left to shade cows in the center of a pasture. As a tree, birds will nest in it. In the south, Spanish moss will grow from its limb. Squirrels will build in its hollows, north or south. Insects, wherever the tree is, will feed on it. Eventually it will die—fall to the ground and in time rot and merge with the soil from which it grew.

Was all this predetermined?

Where does predeterminism end? Or, does it?

We have described the division we imagine existing between ourselves and nature in terms of freedom of choice or capacity to intend. We say we can choose but nature cannot. We say what we do we do intentionally but what nature does is predetermined.

What do we choose? What do we intend?

Do we choose, or intend, to be born?

Do we choose our sex, family, community, nation, intelligence, color, talents, physique, the cultural age in which we are born, this earth?

Do we in fact choose to be human at all?

There are those who assume that such choices are made but even so does this mean that an unborn

Beethoven chose to have music invented, or evolved so that when he was born there would be music?

And what about musical instruments? Did he choose to have the piano invented so it would be there for him to play on and did he choose to have violins, cellos, trumpets, oboes, drums, etc., etc. and all the instruments that make up a symphony orchestra available when he was born?

And did he choose to have paper and ink invented and the means of musical notation that have been evolved over the centuries all ready for him when he came along?

And did he choose to create Bach, Mozart and all their predecessors to evolve a musical tradition into which he would fit?

Clearly, to choose to be Beethoven amounts to choosing to be God, for after all, there had to be a human race, an earth, gravity, an atmosphere, a sun, a moon (for his Moonlight Sonata), forests for him to walk in and of course Vienna, Austria, Europe and everything else that ever has been or will be. So what was it that Beethoven chose? What was left for him to choose or intend once he was born? About the only choice left him was the choice to do the best he could, to do what it was predetermined he was here to do.

Reason would make it appear the only choice we have is that of attitude. Like Beethoven, we can love it or hate it, but love it or not we are here to do what we are here to do and we will do it. The question then is, "How well?"

Here we have located the natural basis for our anxiety, for the question "How well?" is an answerless question. There are no parameters to how well anything can be done. The possibilities of how we can do what we are here to do appear to be infinite. For example, the question of "How well can Beethoven's Moonlight Sonata be played?'' is an answerless question. ''How badly can Beethoven's Moonlight Sonata be played?'' is equally answerless.

Another example is Hitler. Would the UN have been formed without Hitler? Would the state of Israel have been established without Hitler? Would Europe have united as much as it has since World War II without Hitler? These are unanswerable questions: however, they suggest the possibility that on other Earths the Hitler role may be played benignly, that it is conceivable that the Hitler role when played well may be that of a savior. In any event the role of Hitler was played on this earth and whatever he was here to do he did it. Although he did it badly the possibilities for greatness were there and subsequently self-evident.

It would appear our future is predetermined. It would appear we do not choose to intend to be who we are. It would appear we do not choose or intend to do what we are here to do. And yet our anguish tells us we do have a choice or opportunity to intend, to do or not to do the best we can.

Bridge 3

REALITY CRISIS

What we are calling an energy crisis is basically a reality crisis. What we are saying when we say we are running out of energy is we are running out of our concept of reality. To say it or hear it is to believe it, for it focuses our attention on what we know is the real problem. The problem is not the amount of oil or coal left on this earth, nor is it the expense of developing new sources of energy such as atomic energy or solar or wind energy. The problem is our reality is dying. It's all part of what we call our identity crisis. The exhaustion of our fossil fuels is the exhaustion of a past concept of who we are, a passing of concept of reality.

Friedrich Nietzsche (1844-1900) was the first man to say, "God is dead." He was the prophet of the reality crisis we today are calling an energy crisis. To most people the idea that "God is dead" means that no one is in charge. Today most people appear to be as depressed by the idea that God may be dead as they are by the possibility that they may be in charge. This is the nature of our depression over the energy crisis: people appear to fear that no one is in charge but us. This is what a reality crisis is. It's not just a theological or philosophical problem to be worked out within the churches, colleges and certain interested groups. A reality crisis is a crisis of identity, our identity. Putting it simply, reality for us today isn't what it used to be. This is the meaning of Nietzsche's prophecy.

It appears that in our old reality, someone we called God was in charge and we were his servants. Our old identity was that of a servant. According to this earlier concept of reality all that we did on this earth we did because God told us to do it. This concept at its inception was simple and clear. The problem that arose was the idea of "sin." How could we do "bad" if we were doing what God told us to do? "Sin" was the crack in a perfect faith. In assuming there could be "sin" in God's creation man set himself up as the judge of God's works.

The existence of "sin" has been debated for centuries. Regardless of its theological implications, the concept of "sin" marked a new identity for us. By "sin" we meant we were not totally controlled by what we had previously conceived of as the total-controller, God. As time passed our concept of sin enlarged to include responsibility for more and more of our environment. Our sense of being in control broadened. Today there are people claiming that we are not only destroying this earth but are on the verge of destroying the Moon, Mars, and any other planet to which our travels may carry us. In effect they are claiming we potentially possess the power to destroy the universe. They call this "sinful" but what they're really saying is we are in charge, even though they say it regretfully.

It's interesting to note that the pervasive attitude towards our being in control not only of our destiny and the earth's but potentially the destiny of the universe as well, is one of guilt. Possibly it's because of a feeling that somehow we have taken from God something that was not ours to take, but more likely it reflects a general feeling that everything's going wrong and the only one left to blame is us. In other words, we're unhappy that things aren't going the way they used to go.

In general we don't like change and never have. Our way of handling our fear of change initially was to say we weren't responsible for it, God was, and whatever God did had to be right whether we liked it or not. But when we introduced the idea of sin we introduced the idea that we somehow were responsible for change. That's what sin meant. It meant we were assuming responsibility for change. In this fashion all change became suspect of being the product of our sinfulness. For a while we called the changes we learned to like "progress." "Progress" we continued to associate with God's will. However, as what we called "progress" began to produce problems, we withdrew "progress" from God's hands and placed it in our own, claiming as some people do today, that "progress" is our sin. All of which indicates that in the past we accepted all change as a manifestation of God's will while now it appears we are accepting change to be a manifestation of our will.

With this perception it would seem the energy crisis is a crisis of faith in ourselves. But what do we mean by "ourselves"? At the moment we are calling ourselves "consumers." In effect we are saying our function is to consume. Within this concept the energy crisis is a consumer crisis. We are telling ourselves we will have to change our consumer style. In all probability the real change we are facing is that of giving up our identification of ourselves as consumers.

We also see that to think in terms of "who is in control?" is to think in terms of "who is the ultimate consumer?" We see that the relevancy of the idea of "control" depends on the relevancy of the idea of "consumption." We see that the function of controlling is to serve the function of consumption. We see that to think of God in terms of the ultimate controller is to think of God as the ultimate consumer. We see in this context that we have given ourselves two choices: one choice is to serve the ultimate consumer we call God; the other choice is to be the ultimate consumer ourselves.

Clearly consumers are controllers and controllers are consumers. To be one is to be the other. But is this what the cosmos is all about? If we define God in terms of "who is in control?" then this is exactly what we are saying. Our conception of God represents our conception of "what it's all about." In saying God means "control," we are saying it's all about consumption. We are saying God is a consumer, we are consumers, the cosmos is a consumer.

At this point it's interesting to note the Webster's Seventh New Collegiate Dictionary definition of the word "consume" which is "to do away with completely."

The word "consumer" in the same dictionary is defined as "one that consumes: spec: one that utilizes economic goods."

Here we see that accepting the idea of there being an end to the world rests on our acceptance that the purpose and meaning of the cosmos is consumption. Within the context of the Webster's definition of the word "consume" that would mean that the ultimate function of the cosmos is to self-destruct. And this is precisely what is happening if by the word "cosmos" or "world" we mean our concept of reality. Our concept of reality as consumption and control is, in fact, "self-destructing."

This clarifies our understanding of "time." Generally speaking, we only think of time when we are unhappy. Happiness is a state of timelessness. Time is the measure of our anxiety. Realizing the nature of our anxiety, thinking of ourselves as consumers, we fear that one day there won't be anything left to consume. Applying this concept to the cosmos, we say that someday the cosmos will consume itself and that will be the end. And this is how we are talking about the energy crisis. We are saying that within the foreseeable future there won't be anything left for us consumers to consume. And this is what we mean when we say that time is running out for us. We mean that our days as consumers are numbered. By that we think we mean that the fate of the human race will presently be in jeopardy. But it isn't us that's in jeopardy. It's our concept of ourselves; our cosmos, our God, as consumer-controllers, that is in jeopardy.

What makes the energy crisis a reality crisis is that it's a period of transition from one reality to another reality.

Our consumer-controller reality is a "thing" reality. According to Werner Heisenberg, winner of the Nobel Prize in Physics in 1932, "the atom is not a thing." In effect he announced that we are a no-thing part of a no-thing cosmos. Accordingly, we find ourselves moving into a "timeless," "consumerless," "controlless" reality for without a "thing" there is no-thing to control, no-thing to consume, no-thing to measure.

It is for this reason that the energy crisis must be regarded as a reality crisis. At this momentous moment of transition from one reality to another, from one identity to another, all that we have is faith. By faith is meant the act of accepting ourselves as part of a purposeful cosmos without knowing what that purpose may be.

Our age, like past ages of faith, is a time of expanding awareness in which we seek to redefine who we are, what the cosmos is, and what we mean by the word "God."

Bridge 4

CLOSE ENCOUNTERS

Superficially our concern about whether or not UFOs (Unidentified Flying Objects) represent visitations of something like us from other planets appears to reflect our great and age-old concern for whether or not we are alone in the cosmos. On the one hand we fear being alone and on the other there is something flattering about being special. Alternately we react to the thought of other life in the cosmos like an only child torn between wanting to have someone to play with and fearing we will have to share what we have with someone else. So we vacillate between fear and delight at every report that something like us has appeared from outer space somewhere around the earth.

However, lurking beneath the surface of these whimsical fears and delights lies a far greater source of fear and/or delight—the expectation of the advent of our new identity. What we are calling a "close encounter of a third kind," personal contact between one of us and one of "them" from another planet, is a prophetic dramatization of what we are presently calling our "identity crisis."

Historically we have learned that all encounters between different cultures have ultimately eradicated the previous identities of both and out of the synthesis of the cultures a new culture, a new composite identity, emerged. One product of this unrelenting process of synthesis is what is mistakenly called Western culture.

Western culture is neither Western nor Eastern but a culture of synthesis evolved out of the encounters of all cultures on the surface of this earth. The process of cultural evolution is a process of synthesis. By evolution we mean synthesis. Wherever we look we see the process of synthesis occurring. We see it in the process of the growth of a tree. From a seed interacting with soil, water and sun, are synthesized the elements of what we call a tree. We see it watching a cook. Following a recipe, he synthesizes ingredients into a cake or some other form of cuisine. All our arts and sciences clearly evidence various processes of synthesis. The evolution of what we call the human body from a fertilized egg into flesh and blood can only be described as a process of synthesis. In other words, all that we call change is a manifestation of the ubiquitous process of synthesis.

Within this context, depending on how we feel about ourselves, we fear or delight in all change in ourselves as a change in our identity. Most of us hope to become better or fear we may become worse but in both cases there is the acceptance that change means a change in our personal identity. Most of us either fear or hope that our identity is not a constant. Given the apparent fact we are, whatever that may mean, part of a process of synthesis we are right. That we aren't the person we used to be is witnessed by any family album. That one day we will not be what we are today is evidenced by graveyards. Still our hopes and fears seem to center on what we call our loss or change of what we call our identity as though we had an identity that was ours to lose or change. It's important therefore to try to understand what we mean by "identity."

To begin with the obvious, we are what we think we are. This is an irrefutable statement, for even to say, "I am not what I think I am," is simply to suggest that I can't think of what I am at the moment but given time it will come to me. On reflection it appears it is virtually impossible to think of ourself as something we can't think of. This tells us we think we are a "thought." Whatever the thought may be does not alter the fact that it is a thought, although the variety may be almost infinite.

Upon saying this it becomes equally clear that what we call "reality" is what we *think* it is. This makes "reality" a thought. To say this is to say that everything is a thought. Within this context, evolution, the process of synthesis, appears to be a thought process. Also within this context whatever answer we may choose to give to the question "whose thought?" is in fact also a thought. Any answer to the question "can a thought think?" will, of course, be another thought. If then, we mean by the word "consciousness" our thought process, it follows that what we call the cosmos is our consciousness. To put it another way, if thoughts are words, we, the cosmos, all that we call "reality," are composed of words. In other words, to be conscious is to be verbal. What we are then saying is that to be is to be verbal. We may therefore conclude that we are part of a capacity to articulate words. We are currently calling this capacity "mind" but traditionally we have called it "God." According to the New Testament (John

1:1), "In the beginning was the word." Clearly the words "God" and/or "Mind" mark the line between the verbal and the non-verbal, being and non-being, between consciousness and non-consciousness.

Do they also mark the line between feeling and non-feeling? If so the line is not as clearly drawn as we might imagine for we still, though highly verbal, do feel. Perhaps then the line indicates a shift in emphasis. Perhaps before words the emphasis was on feelings and perhaps within this verbal phase the emphasis is on words.

If, as some people say, we are moving from a verbal identity towards a numerical identity, that the reality of the future will be a mathematical reality; if such a transition occurs, we may reasonably expect that the emphasis on feelings will one day be minimal. Indeed, that feelings may ultimately be phased out entirely. But regardless of the precise nature of our next identity, whether it prove to be numerical or what may be more likely, a synthesis of numbers and words, history records a steady decline in our confidence in feelings. We have become increasingly distrustful of our emotions. We have recently made a science of this process of disassociation from our emotions. We have called it *psychology*. The function of the practioners of this science is to help patients deal with their emotions. In effect, they are saying that we and our emotions are separate entities, in the same fashion that medical doctors had previously taught patients that we and our bodies were separate entities. In both cases a line has been drawn between the "real you" and a "non you." In the case of psychology the message has been that you are not your emotions—that is, you are not what you feel. In the case of medical science the message for some time has been, "You are not your body." We are beginning to regard our emotions in the same light that we regard bacteria, as something alien, out to get us.

Recently, however, medical science has discovered that bacteria is not an enemy. Apparently what "gets" us is not bacteria but an over-reaction by our bodies' defense systems. As Dr. Lewis Thomas puts it in his book *The Lives of a Cell* (pp. 92), "Our arsenals for fighting off bacteria are so powerful and involve so many different defense mechanisms, that we are in more danger from them than from the invaders. We live in the midst of explosive devices; we are mined." The primary triggering device for our bodies' defense system is our nervous system, which in turn is triggered by our emotions. Not only are we no longer to identify with our emotions, we are now regarding them as potentially our greatest enemy. And this is true on most if not all levels of present-day life. For example, there is great concern about the emotional state of anyone in a position to press any buttons capable of firing rockets with nuclear warheads. The ever-growing proliferation of agencies committed to the protection of children, wives, elderly people, minorities, not to speak of birds, fish and game, from the effect of violent emotions is another example of our growing fear of our emotions. As is the growing use of drugs such as tranquilizers, alcohol, marijuana and cocaine. Contemporary man is deliberately doing all he can to contain, pacify, if not stamp out once and for all, what he calls emotions.

Then there is television. TV is the ultimate process abstracting us from our visual identity, which presumably was our first form of personal identity. Until the moment that the first one of us saw his or her reflected image in a still pool of water all identity must have been collective. We must have thought of ourselves as simply one of our particular group—as monkeys and tigers are included in the body of their specific groups. But what happened when we saw and finally accepted what we saw as our specific image was the inception of the concept of a personal identity. Our first concept of the individual was a reflected image in a still pool of water. At that point our reflected image was ours and ours alone but when the first artist drew the first portrait this ceased to be. The artist followed by the photographer not only took our visual image from us but also took our visual reality from us. Our visual identity and our visual reality do not belong to us but belong to artists and photographers. But only the first artist and to a lesser extent the first photographer could claim that their visualizations were theirs and theirs alone for all subsequent artists and photographers have drawn their views of visualized reality from their predecessors. The evolution of the visualization of reality has been, as all evolution is, a process of synthesis. The ultimate act of the democratization of our visualized reality has been performed by television. The television camera has given our visualization of our selves and the reality to the world and to the cosmos. Even now our televised images are radiating out from the earth to be seen on any television screen anywhere capable of receiving them.

Thus stripped of our visual and emotional identity, momentarily accepting a verbal identity, we wait alternately fearing and joyfully anticipating the advent of a new identity, presently conceiving of this momentous event as an encounter of a third kind.

In the acceptance of ourselves as a thought, regardless of the nature of the thought, whether it be verbal, or mathematical, we are preparing for such an encounter. We aren't looking for a new visual identity, nor a new emotional identity. What we are looking for is a new idea of who we are, a thought we have never thought before. Ultimately there will synthesize out of such an encounter a thought neither party has ever thought before and that thought will mark the beginning of a whole new species of thought.

As we watch for UFOs, we are longing for or prophesying a new identity for man.

Bridge 5

NATURE

At some point in history we conceived of ourself as separate from the cosmos—an orderly, harmonious, systematic universe encompassing everything in nature. We drew an imaginary circle proclaiming that all that was inside the circle was us and what we made, and all that was outside the circle was nature and what nature made. This led us to accept ourselves and all that we made as non-natural.

At present the popular view is that "nature" is good and we, being non-natural, are bad. In his essay *Origin of the Inequality of Man*, Jean-Jacques Rousseau didn't go that far. Rousseau simply argued that we were naturally good; it was our culture that was bad. In other words we weren't bad, it was what we made that was bad. Rousseau (1717-1778) wrote this in the late seventeen hundreds.

In *Heart of Darkness* published in 1902 Joseph Conrad stated the case eloquently for our Victorian view of ourselves. The reverse of Rousseau's, our Victorian view was that it was the culture we evolved that was good, and that those of us outside our culture were savages. As Victorians it was nature that we judged as evil and we feared it. We felt that within ourselves there were still the dark roots of nature which in an unguarded moment would take over and we would revert to naturally bestial behavior. The character in *Heart of Darkness* who faced this beast in himself is called Kurtz. The most memorable line in the book is when Kurtz attempts to express his reaction to this confrontation: "The horror! The horror!" he cried.

In an earlier a more primitive time nature appeared to us as manifestations of spirits—that is trees, fish, birds and animals. The elements and the heavenly bodies represented gods. The sea was a god as was the earth, the moon and the stars. For us then, survival appeared to depend on how well we got along with nature, the spirits, and the many gods.

Then as Hebrews we conceived of ourselves as the chosen ones and all nature as here to serve us. In effect, this concept reflected a confidence in our capacity to survive. As Hebrews we said what had long been in our hearts, that our relationship to nature was that of an adversary. We had always feared nature but as Hebrews we finally said it out loud. Our way of articulating this proclamation was to strip the world and all heavenly bodies of gods, claiming there was one god and that god was in our image. To put it the way we put it then, we said we were made in the image of the one and only god. The effect of this statement was to give us new confidence in our being and our destiny. This proclamation marked the most momentous identity shift that has occurred in our history as humankind. In one stroke we reconceived of ourself—changed our view of ourself from that of a victim of nature to its master.

Still, even with the Hebrews the adversary relationship of ourselves and nature held. Now this cyclic attitude of ourselves towards ourselves and nature is questioned. Our relationship with nature is not an adversary one. We and nature are all part of one process.

Take a molecular view of the world. What do we see? Everything in nature including humans is composed of molecular processes composed of atomic processes. All molecular processes are comparable. All atomic processes are comparable. The difference between humans and nature is not in that one is composed of molecular-atomic processes and the other is not. Both humans and nature are composed of molecular-atomic processes. In other words, we and nature are made of the same stuff, or processes, but the molecular and atomic processes are in different arrangements or patterns.

This holds equally true of what we make and what nature makes. The difference between what we make and what nature makes is the difference between "brands." We may prefer one brand over another but both brands are made of the same stuff. There is nature's brand and our brand and it really is as simple as that.

Take a television screen and split it down the middle. On one side we see a time lapse series of pictures of grass, flowers and trees growing. On the other side we see a time lapse series of pictures of a city growing.

What's the difference?

In both pictures we begin with the surface of the earth and in both pictures we see something growing up out of the earth. Cities grow up out of the earth. The wood, the sand, the water, the ores that go into growing grass, flower and trees are the wood, sand, water, ores that go into growing cities. In fact cities grow up out of the earth in clusters very much like mushrooms do.

Take another twin TV sequence of time lapse pictures. On one half see a film of the evolution of a river. Begin with a small spring pouring water out onto the land, gradually coursing downhill, cutting into the soil and creating a bed for itself as it goes. In time we see other streams joining it until finally they all form a great body of water that in turn carves a large river bed out of the soil of the valleys.

On the other half of the screen see the evolution of a railroad. Just like the stream, the railroad cuts a bed for itself through the valleys and joins other rail lines until finally a whole river of rails runs into a city.

Now, put away the split TV screen, and in its stead put a camera in orbit around a large ball of gas in space (composed of molecules composed of the processes called atoms) and record a time lapse series of two acts in the evolution of the earth. As the time lapse series unfolds see the condensation of the ball of gas into a revolving rock wrapped in clouds. The clouds obstruct our view for a bit but finally through clear patches we see a blue planet. The surface is heaving up; mountains and torrents of rain are filling huge cavities in the surface to form seas fed by cascading rivers. Splotches of green appear here and there, spreading out over the land masses. A closer view shows the green is composed of grasses and large asparagus-like trees. Moving about in this greenery are the animals we have come to call dinosaurs. Then, from each pole great white glacial masses of ice and snow descend like curtains and end one act. They recede and a whole new act begins. Gone are the asparagus-like trees and dinosaurs and where they were there now are the forebears of the trees and animals that currently occupy the surface of the earth.

Now think back a moment: in among the asparagus-like trees and dinosaurs something was scurrying about. It was very small but we recognized it as one of our earliest ancestors. It, too, is gone as the curtain of ice and snow recedes, and in its stead is something that looks a lot more like us than its predecessor. In time it comes to look exactly like us. It is at this point that this thing becomes truly interesting. With surprising suddenness it begins to move, taking the elements of the surface of the earth and resynthesizing them into cities, paintings, musical instruments, sculptures. Eventually it takes rocks and makes them fly. That is, the creature, us, converts ores into metals. These metals, shaped in particular ways and fueled with energy banked in the earth, by prehistoric plankton and plants, are capable of moving through the atmosphere many times faster than the speed of sound.

But then the flying rock does something nothing on earth has ever done: leaves the earth and lands on the moon.

Has our camera recorded a battle between nature and humans? No. What have we seen? We have seen a process that began with a gas cloud condensing into a ball that became an earth and from that earth there now has flown a what? Well, what should we call it? What's it like? It's like a seed pod. The earth has tossed out a seed pod and within that pod is a form of life that is capable of growing on the moon and probably a lot of other planets we haven't set foot on yet.

So there we are. We can either say that we are working for nature, or nature is working for us, but what we cannot say is that we and nature are adversaries for we obviously are not. From this perspective we and nature are not adversaries but part of a process that is part of a process that is, etc., etc., etc.

Bridge 6

MAN AS DNA

Whether there is or is not an evolutionary plan for a cultural body and if there is, where or what the plan is, has been a matter of debate. It is becoming clear there is an evolutionary plan which we have been following in evolving our cultural body. The plan appears to be our biological bodies. The relationship between our biological body and the evolution of the "cultural-body-of-humankind" is comparable to the relationship between DNA (deoxyribonucleic acid) and the evolution of our biological body. Our biological body and DNA both appear to be plans. When building our biological body each cell has a copy of the overall plan and constructs a small part of what is needed. Just so, each of us builds some small part of what is needed for the greater body, the evolving cultural-body-of-humankind.

In observing the cellular evolution of our bodies we see a single fertilized cell divide into two comparable cells. Each cell then repeats this process. All these cells have comparable DNA. These cells begin to form into different groupings, building different bodies we call organs, like the liver, heart, stomach, kidneys.

In the evolution of the cultural-body-of-humankind we see a parallel process. Just as cells multiply, people propagate over the centuries and gradually break up into different groupings of color and language, forming what we have come to call nations, each nation an essential organ of the evolving cultural-body-of-humankind. Until recently we didn't have a way of understanding these events. We lacked what we call a model.

The discovery by Watson, Crick and Wilkens in 1953 of how the DNA molecule functions provided that model. DNA is the key to understanding how we are evolving the cultural-body-of-humankind. The cells in meeting needs of DNA—their blueprint for building—build our biological bodies. Our biological bodies in meeting their needs are building the cultural-body-of-humankind.

We have been following body-plans in the evolution of the cultural-body-of-humankind. Our body plans are our appetites. In serving our bodily needs we have evolved and are evolving the cultural-body-of-humankind. The motivation for the evolution of our cultural-body-of-humankind has been to build a better way of life. Who asked for a "better way of life"? Our bodies.

Looking backward from the edge of the 21st century we notice that the first thing we did on earth was look for food and water. How did we know we needed food and water? Our bodies told us. Obeying our bodies' appetites we discovered all kinds of food. We also discovered all kinds of things to drink.

For the same reason that we looked for food and water we also looked for shelter.

With the ice ages we found ourselves looking for clothing. The only available clothing was that worn by animals. We learned how to skin animals. In effect, we were taking ready-made clothes off the rack. In the process we evolved ways of skinning animals and tanning hides. Our bodies told us how to do what we had to do just as DNA tells the cells how to do what they must do. Eventually we discovered how to make our own clothes, grow our own food and build our own shelters. Villages came into being and evolved into cities. Because our bodies demanded heat we discovered how to use fire. We have evolved the industrial-electronic age because our bodies want it.

Until we found food, water, shelter and clothing we didn't know there were such things as food, water, shelter and clothing. We never knew there was such a thing as an industrial-electronic age until we evolved it. Our bodies appear to tell us first to look for something we've never heard of, then tells us when we've found it, or built it. Then we name it.

Every time we make a discovery or find a new way of satisfying our bodily needs our bodily needs increase. It would be reasonable to expect that once we satisfied a bodily need that would be that. That isn't what happens. What happens is with each solution our needs diversify and multiply. We have deplored our insatiability.

But in serving the needs of our biological bodies we have not been self-indulgent. We didn't make up the need messages we have been receiving any more than the cells made up the need messages they receive from DNA. In meeting these needs we have been fulfilling our natural function on this earth, that of building a cultural body. Entertaining the idea we thought the whole thing up, and plagued by the fear culture was happening in a hit or miss, random fashion, we have been unaware that from the beginning we have been following a plan.

With DNA we discovered that not only human life but all life evolves according to a plan. All life has DNA. This means all life evolves according to specific sets of needs we call DNA.

Animals, birds, reptiles and fish all have specific needs. All forms of vegetation have specific needs. Each form of life appears to satisfy the needs of other forms of life creating an environment of mutual need.

Directly or indirectly all forms of life need sunlight. Sunlight is not considered to be a form of life. Water is not thought of as a form of life, but without water there would be no life on this earth. All forms of life directly or indirectly need air, the atmosphere that envelops the earth. Air is not considered to be a form of life. All forms of life need minerals but minerals are not considered to be a form of life. How can life need non-life? Yet without non-life there would be no life on earth.

The question that arises is, is there a mutual need between what we call life and what we call non-life, as there is between all forms of life? Does what we call non-life need what we call life? Does sunlight need what we call "life"? Do water, air and minerals need what we call "life"? Does the earth need life? Is there a DNA that articulates the needs of these forms of non-life?

DNA is a molecule. A molecule as defined by Webster's New Collegiate Dictionary is: "1: the smallest particle of a substance that retains the properties of the substance and is composed of one or more atoms." A molecule is not considered to be a form of life.

The needs of the molecule DNA, a form of non-life, instruct living cells how to build living bodies. The atoms that form DNA must have united under a comparable compulsion of need. The molecule DNA must itself be built according to a comparable need that instructed the atoms how to build it.

No one has ever seen an atom. Werner Heisenberg, 1932 winner of the Nobel Prize, said in his autobiography, "Atoms are not things but processes." This atomic process might be described as a manifestation of need. The bodies that compose the cosmos are said to be composed of atoms. If so and if the atomic process is indeed the manifestation of need, then the basic ingredient of the cosmos could be said to be need. The function of our cosmos would then be to need.

What does it mean to need? According to Webster's New Collegiate Dictionary to need is: "1: to be in want 2: to be needful or necesssary: to be in need of: REQUIRE: be under necessity or obligation to (he need not answer)."

If our cosmos is a manifestation of need, evolution appears to be a ceaseless attempt to satisfy an insatiable need. What does our cosmos need?

If as Werner Heisenberg says, "The atom is not a thing but a process," and we say that process is need, is our cosmic need for need? Put another way, is our cosmic need to be needed? Who or what then needs our cosmos? It is a metaphysical question requiring a metaphysical answer.

"Metaphysical" as defined by Webster's New Collegiate Dictionary means: "1: of or relating to metaphysics 2a: of or relating to the transcendent or supersensible."

According to Webster's New Collegiate Dictionary "transcendent" means: "1a: exceeding usual limits: SURPASSING b: extending or lying beyond the limits of ordinary experience c: Kantianism: being beyond the limits of all possible experience and knowledge."

"Supersensible" as defined by Webster's New Collegiate Dictionary is: "being above or beyond that which is apparent to the sense: SPIRITUAL."

In other words the insatiable need our cosmos seeks to satisfy is beyond our understanding. The nature of that need, as postulated by Webster's New Collegiate Dictionary's definition of "supersensible," could be spiritual.

"Spiritual" is defined by Webster's New Collegiate Dictionary as: "1: of, relating to or consisting of spirit: INCORPOREAL (man's spiritual needs)." "Spirit" as defined by the same dictionary is: "1: an animating or vital principle held to give life to physical organisms."

If we identify "life" with a capacity to need it would appear there is no non-life for need appears to animate our cosmos.

The nature of this need appears to be instructive.

The needs of DNA instruct cells and with these instructions cells build biological bodies. The needs of our biological bodies teach us how to evolve culture. The specific needs of specific atoms presumably teach them how to evolve DNA molecules. It would seem reasonable to assume that the collective needs of air, sunlight and water instruct the collective biological systems on earth teaching them how to evolve ecological systems. Earth's ecological systems reflect the mutual needs of sunshine, air and water for biological systems.

The mutually instructive needs that bind the earth and sun Sir Isaac Newton defined as gravity. But the word "gravity" implies a mechanical relationship. It would appear that the relationship may be better described as a mutually instructive relationship. How the earth instructs the sun is not clear but how the sun instructs the earth appears to be evident in the orbital course the earth has learned to take in circling the sun. The fact that the earth is so angled in relation to the sun as to have seasonal changes is another indication of how the sun might instruct the earth.

Needs are mutually instructive.

Somehow cellular needs for the needs of DNA are mutually instructive. Somehow DNA's needs are mutually instructive with those of the atoms that compose DNA. Somehow earth's gravitational need to hold its ecological systems is matched by the ecological systems' mutually instructive need for it.

Humans are today being pulled off this earth by a need greater than the pull of earth's gravity. That need would appear to be the seeding of life on other planets. It appears other planets need us. That need is in the process of teaching us how to build space ships, space colonies and how to colonize other planets. We are being instructed by planets that need us for something we don't yet understand just as we initiated the industrial-electronic age without envisaging the outcome.

It must be assumed a mutually instructive need exists between us and what we call culture. Viewing ourselves as cells we see we are building a cultural body. This cultural body left earth before our biological bodies, broadcast into the universe first by radio, then by television. In this context earth appears as a womb. We appear to be cells that in that earth-womb built what we call a cultural body. That cultural body is now in the process of being born into the greater universe. The reciprocally instructive need between our cells and our bodies is evident. Such reciprocal need must exist between our biological bodies and our cultural body, as it must exist between our cultural body and the cosmos.

It must also be assumed a mutually instructive need exists between humans. Each and every human must need and be needed. Each of us is needed to instruct the other.

Each and everything in our cosmos must need and be needed. What is called interaction or feedback is mutual instruction. The force that animates our cosmos appears to be mutually instructive need.

Perhaps what we mean by spirit is a mutually instructive need, its purpose unfathomable in its intent.

However obscure the cosmic intent, man can now

know something about himself he never knew before. It has to do with the purpose of his leaving the surface of the earth and his predictable effect on the solar system. There is no question he evolved the earth into something it was not before his existence on it. However minutely, the Moon and Mars have already been changed by his presence on their surfaces in one form or another. There is talk of mining the asteroids in the asteroid belt. Clearly, man is an instrument of change.

He, it appears, is the DNA of the next step in the evolution of the solar system.

Bridge 7

RECOMBINANT DNA

We may shudder at the possibility that Recombinant DNA may create a monster comparable to the one created by Mary Shelley's Victor Frankenstein. But the production of a new species, many of us are beginning to suspect, may be our natural function, a shift from an earlier perception of our true function as the production of goods and services.

It is clear that with the evolution of cybernation and automation, we are evolving rapidly out of all repetitious work. It is clear that work that can be repeated is repeated more quickly and efficiently by machines operated by computers. These are the jobs that produce "goods" and "services" and we are in the process of working ourselves out of them. This is something that most know and many are concerned about. The question being raised is, "Is man no longer useful?"

This sounds like a slave recently released from a lifetime of forced labor complaining, "What will I do now?" Historically our function has appeared to be clothing, feeding and sheltering ourselves. Until now we have been virtually slaves of these needs of our biological body, and we have resented this function. The resentment has been a constant. It warrants serious consideration as an indication that our true function is something other than work to maintain our body.

Our way of describing our other function appears in our conception of religion. Out of this conception we evolved what we call the arts and the sciences. In these disciplines we see the continuous realization of our ever-new images of ourselves and what we call God. Both arts and sciences deal with our perception of a cosmic purpose and our relation to it: our arts attempt to describe an assumed cosmic purpose; our sciences attempt to explain how the cosmos works.

These two distinct functions reveal our evolution in two distinct aspects. In its first aspect we see that in laboring to provide food, clothing and shelter for the survival of our body we are behaving like all other forms of life on this earth. In its second aspect we see that in our conception of religion and the subsequent evolution of the arts and sciences we are doing something, so far as we know, no other form of life on this earth has done or is doing. The difference between these two aspects of our behavior is that one is designed to serve the needs of our biological body and the other is not. (It is true that our sciences subsequently spawned a bevy of applied sciences designed to serve our body but this is not the initial motivation of science. Its initial motivation was to discover a cosmic purpose.)

Why have we—mankind—spent so much time and effort on the production of something we can't eat, wear or use for shelter? There have been as many answers to this question as there have been artists, scientists, philosophers and priests. All the answers have been different but they all have agreed on one point: the function of religion and arts and sciences has been to serve not our bodies but something none of us has ever seen. Something all of us have felt. Something that tells us what to do. It is something we call soul or mind. They also agreed in varying degrees that the needs of this something we call soul or mind are of greater importance than the needs of our bodies. Throughout our history most of us, most of the time, have agreed with them.

It would seem we have been serving two masters, a body and a mind or soul. In the service of our bodily needs we evolved the industrial revolution. In the service of our mind or soul needs we evolved what we call religion—arts and sciences—culture.

At the same time that developing automation suggests the possibility exists we may no longer be employed in the jobs that feed, clothe and shelter our bodies, we are faced with the possibility of doing something we may have dreamed of doing but never expected to do—build a better body than the one we have on. Suspecting that we can do it, we agonize over whether we should. The answer is we are already trying to do it.

Whatever we have thought we have been doing in this industrial-electronic age, we have in fact been building extensions to our biological body. To realize this, think of being divorced from cars, trains, ships, airplanes and our world of electronic communications. These are called conveniences but without them we would be a different people, living different lives, unable to do most of what we presently are capable of doing. We would be disembodied from the technology that has become our greater body.

One of the advantages of our greater body is all its parts are replaceable. Replaceability has proven to be such a great advantage that there are very few mechanics or electricians left. What's the point of trying to repair what can be replaced more easily and generally more cheaply. Now we are applying the idea of replaceability to our biological bodies. Crutches and wooden legs marked the crude beginnings for a highly sophisticated industry concerned with the development and manufacture of spare parts for the biological body. In laboratories around the world research and development teams are working on replaceable hearts with replaceable parts, television eyes for the blind, artificial limbs with electronic nervous systems that "feel" and more.

Some research scientists along with many science fiction writers are already asking, "Why keep replacing parts to this body? Why should we serve its needs? Why not design a body that serves our needs?" By "our needs" they mean those of the soul or mind.

Why not go back to the drawing board and design a better body?

What could they mean by a "better body"? Possibly the solution to our economic problems rests not in the design of a more efficient economy but in the design of a more efficient body? The experience of poverty has its roots in the design of our present body. This body must continuously be clothed, fed and sheltered. Why not design a body that doesn't get sick? A body that doesn't break easily? A body more adaptable to living on other planets than planet earth?

Until now we have accepted our bodies as constants and their environment as a possible variable. Today we are beginning to consider both our bodies and its environment as variables.

The first step in the direction of designing a new body is being taken with the process called Recombinant DNA. Biologists are now mastering the process of taking genes from one organism and splicing them into another organism. A whole new industry is evolving around the employment of Recombinant technology. At present we are seeking ways to use this technology to build better babies. When a faulty gene is detected in an unborn child that faulty gene may be replaced by a healthy gene. It would seem to be only a matter of time before we discover ways of making our own genes as opposed to simply inheriting them. We will then be in the business of designing a new species.

We are body-builders naturally. Throughout our history on this earth we have been building bodies. We built our own greater-body—cities, communications networks and factories. What would be more natural than that we build individual bodies for a new species? Perhaps our true function on this earth is parental, to conceive the next species of humans. A species capable of migrating more easily from this earth into the greater universe. It might be that we are DNA for the next species of humans. Where did DNA, the plan for our present biological bodies, come from? For that matter, where did the plans for our greater-body come from? Trace the evolution of the airplane back as far as birds and on, as far as you can, and you still have to ask, "Where did that plan come from?" None of us has ever done anything we thought of doing without a plan. We didn't, for example, make up our own language. Some say the roots of our language go back to our bodies' DNA. And where are the roots of DNA? As far as we know they extend back to the stars that, so far as we know, condensed out of hydrogen clouds billions of miles long and billions of miles wide. And where did the plan for those hydrogen clouds come from? For that matter, where did the plans for the hydrogen atom come from?

It would seem that whatever we do, we do naturally. We have arrived where we are by listening naturally to a voice within. If we do, as it now seems probable that we will, design a new species of humans it will be natural for us to do it. We need not shudder.

Bridge 8

CRIME

There is a myth that to pass a law is to eliminate a crime. In fact, to pass a law is to create a crime. The simple truth appears to be that laws increase the number of crimes and the size of the criminal population.

According to Webster's Seventh Collegiate Dictionary, "Crime [is] an act or the commission of an act that is forbidden or the omission of a duty that is commanded by a public law and makes the offender liable to punishment by that law esp: gross violation of a law." This suggests the quickest, easiest and least expensive way of eliminating crime is to eliminate laws.

According to Webster's definition, laws define criminals; hence, without laws there can be no criminals. The first criminal was created when the first society created the first law. Before that it would follow the world was crime-free, as the Bible says it was at the time of the Garden of Eden, or in the time of Jean-Jacques Rousseau's "natural man." It is equally apparent, according to this definition, that with each new law a new criminal category is created. What we see here is what appears to be a direct correlation between a rise in the number of laws passed and enforced and a rise in the crime rate.

The word "enforced" is important, for the more laws that are enforced the higher the crime rate.

Or so it seems, but on closer inspection we might discover that what in fact is being recorded are rising and falling efforts to enforce the laws.

However, going back to Webster's definition of "crime," it does appear legislatures create criminals. It might therefore be a wise precaution to demand that each new law offered for passage include a projection of the possible number of new criminals this new law will create. That would allow the law enforcement agencies to request the additional funds considered requisite for the enforcement of the new law. Shouldn't it be mandatory that before the passage of a law hearings be held at which law enforcement agencies are requested to give their estimates of the cost of the enforcement of that law? Thus with the passage of each new law appropriations could be made for an appropriate increase in the number of judges in the land and the addition of more prison cells to accommodate the legislated increase in the criminal population.

After all there is no point in passing a law if there is no intention of enforcing it and the tax payers should know approximately how much enforcement is going to cost them.

If this were done the rising cost of law enforcement, if it kept pace with the numbers of laws passed, would in all probability exceed the cost of sustaining our national defense system. It's doubtful if the nation can now afford to enforce the laws it already has passed.

Today in any society called civilized it is almost impossible for any of its members to go through a day without breaking at least one law. Ignorance of the law is not legally acceptable as an excuse but civilized societies have passed more laws than any one judge or lawyer, let alone layman, can know or remember. To be a member of a civilized society is to be a member of a society largely composed of criminals most of whom have not so far, knowingly or otherwise, committed what Webster's Dictionary describes as a "gross violation of a law."

It's shocking to realize but it appears to be true that to be called "civilized" is to be what is called a "criminal." A curious state of affairs.

Is the solution to do away with laws? If so the next question would seem to be, "Can society survive without laws?" But neither one of these questions gets to the heart of the matter. Let's ask the obvious question: "Can society survive without criminals?"

If the answer to this question is "yes" all we have to do is discard the concept of "criminal" and reconsider what we mean by "law." If the function of the law is not the creation of criminals, what is its function?

Interestingly enough there is already a move to reconsider what we mean by law. There are now what are called "no-fault laws." These laws mark a significant shift in emphasis from fault finding to seeking an equitable solution to a problem. In states that have no-fault laws pertaining to traffic accidents the intent is not to create a criminal but to care for the injured parties. In states that have no-fault laws pertaining to divorce the intent is not to prove infidelity as the basis for separation but to prove incompatability. This shift in our concept of the function of law is possible because of a change that is occurring in our thinking about ourselves. We are changing from thinking of ourselves as all members of the family of man to thinking of ourselves as all members of one body, the cultural body of humankind.

There is a big difference between these two perceptions. For example, most homicides occur within the family. What is called *homicide* within a family is called *suicide* within one body. While we might be prepared to murder or dismember a member of our family we aren't in as great a hurry to cut off our own legs, poke our eyes out, or cut our heart out. Within the context of a family "punishment" may appear to be rational, but within the context of one body "punishment" amounts to self-abuse, a form of irrational behavior.

Within the context of a family all problems might be dealt with in terms of "who did it?" with the intent of blaming and punishing someone or someones. Within the context of the body all problems are problems of healing. Adversary relationships appear to be rational in a family context while they appear to be irrational within a one-body context.

So long as we think of ourselves as members of one family we can think in terms of "good guys" and "bad guys." The function of the law in such a context is to define the bad guys as criminals. This criminalization of individuals, and in some cases whole segments of society, allows the state legally to abuse its own citizens.

The past concept of all people as members of one family applied on a world scale led to the acceptance of nations as branches of that one family. This led to the evolution of what is called international law. The concept of the function of international law has proved to be identical to that of the function of domestic law, from which it naturally evolved. It has defined who are the good guys and who are the bad guys. The process of international law so far has been the criminalization of nations. This effect legally allows one nation or group of nations to abuse another nation or group of nations. Within the context of the family of man, war is tolerable or at least conceivable

With the concept of all people as members of one body, the cultural-body-of-humankind, the word "war" becomes irrelevant, meaningless, and is replaced by the relevant, meaningful word "suicide." The same holds for the word "crime." The word "crime" is simply a meaningless noise within the concept of humankind as one cultural body. What within the family of man we called crime, within the cultural body of humankind becomes a problem that requires healing.

Within our new consciousness there is no concept of crime. This shift of consciousness eliminates all crime both national and international. It makes even the idea of crime unthinkable.

Bridge 9

EXODUS

We are outfitting our houses with increasingly sophisticated nervous systems. The components of these nervous systems we call door bells, fire and crime detector warning systems, thermostats, telephones capable of answering themselves, radios capable of turning themselves on and off at specific times, and even capable of turning on appliances such as coffee makers, etc. Other components of our houses' nervous systems are television sets with memory banks to record selected shows at selected times, television cameras to monitor sleeping children and sick people or whoever is at the front or back doors. And there are lights that turn on at dusk and off at dawn.

Our houses are becoming much more like our human bodies.

In the beginning a house was only a shelter, a shell into which we could crawl to seek protection from rain, snow and predators. With the discovery of fire we added heat. At this point the house became, like us, a consumer of energy. With heat came cooking. Cooking may be defined as a digestive process, so it could be said that with the advent of cooking our houses, like us, ingested food. With plumbing we gave the house something like a circulation system plus an alimentary canal. Windows act as pores, doors act as mouths, and perhaps in the case of rear doors through which the trash passes, as anuses.

In the evolution of our houses we apparently have been and still are evolving bodies for our bodies to live in.

Looking back over the process of our cultural evolution it becomes clear that we are a "body-builder." In early wars, for example, we evolved a body that gradually encrusted our original body. We called it armor and it amounted to a kind of skin less capable of being cut or bruised than our flesh skin. Subsequently we evolved faster, stronger, larger bodies such as seagoing battleships and land-battleships we called tanks, followed by airplanes.

Daily, the average person occupies a number of bodies other than his own, beginning with a house, followed by a car, then perhaps a bus or commuter or subway train, then an office building or factory and occasionally an airplane or ship.

On the macroscopic level the primary difference between these bodies and the human body is one of complexity.

On the microscopic level these bodies are all composed of comparable molecules, the only apparent difference being in the molecular arrangements.

On the sub-atomic level all differences disappear.

What all human bodies have in common is they die. We have always feared death. Concurrently with our development of bodies for our bodies to occupy we have been progressively evolving various paths of exodus from our original fallible bodies. The first faltering efforts in this direction occurred in our evolution of words, numbers and music.

Intuitively we seem to have felt we could move from the confines of present bodies to that of other bodies in the form of thought patterns. What this indicates is that at some point we began no longer to identify ourselves with our bodies and began to identify ourselves with our thoughts, or thought patterns. It is not known exactly when this occurred but it marks the most momentous event in our history.

At first the extent of this form of exodus or transfer from dying bodies was limited primarily to contemporaries. Some few remembered thought patterns were, however, passed by word of mouth from one generation to the next. This form of exodus introduced the staggering possibility of a superhighway to eternity for our thought patterns.

With the evolution of the means of notation of words, numbers and eventually musical notes, it became evident all verbal, numerical and musical thought patterns could travel this superhighway.

Beginning with cave drawings and carvings, through the evolution of painting, sculpture, photography, the movies and now television, with holograms now in an advanced state of development, we have ceaselessly evolved the means of translating our self into thought patterns.

Thus the medium of this superhighway for thought patterns was first the human voice, followed by stone, followed by paper, canvas, film and magnetic tape to be followed presently by the laser beam.

From our first crude drawing of our body to the evolution of medical science we have sought to understand our body down to its smallest detail. For it is by this process of "understanding" that we translate ourselves into words, numbers and musical notes. The function of "understanding" appears to be the process of self-translation into a communicable thought pattern.

We now know that our body is quite capable of curing itself but this awareness came to us through the evolution of medical capacities that allowed us to see how our body works. Once we have seen all there is to see, the need for medicine will give way to the cure of clear thinking. What we are doing is translating our body into thoughts. Once we have accepted our bodies as a thought composed of thoughts, we can no longer accept it as a thing to be dealt with by things, but as a thought to be thought clearly.

When Werner Heisenberg, the winner of the Nobel Prize in Physics in 1932, discovered that the atom was not a thing but a process, we realized that what we called bodies were not things, composed of things but processes composed of processes. Since we think in terms of thoughts, we naturally must assume that by "processes" he could only mean thought processes. If everything was made of "no-thing" what else could it be made of but thoughts?

In following our escape route from our body through the process of translating our self into thought patterns, we have been led into the world of the media. By the world of the media, we mean the world of thoughts. When we complain of the media we are complaining about our thoughts.

What we have traced in our exodus from our bodies, is the process of our conversion of our bodies into what we currently call the media. Or, to state it more precisely, we have traced the translation of one language into another language. What we have traced is the translation of the language of our bodies into the language of words, numbers and musical notes.

Seen in this context our bodies may be viewed not as "things" but as vocabularies. Take, for example, the fertilized cell from which what we call "life" evolves. This cell proceeds to divide itself into two cells that then divide themselves, each, into two more cells, and so it goes—each subsequent cell subdividing until a critical number is reached. Then the apparently identical cells all commence to perform unique functions that in unison compose what we call a body. It is as though there first must be the articulation of a vocabulary of a certain size before the words can be assembled to form sentences, as in the case of simple organisms, and paragraphs as in the case of more complex organisms. In the case of highly complex organisms, whole books. In all probability the cells are far from identical but, being part of the same vocabulary, may look alike to our untutored eye.

Individual people look and act very much alike but when enough assemble to form a community they begin to act differently. When enough assemble to form a city the differences in behavior are greater. When enough assemble to form a nation the diversity in behavior patterns is even greater until, as today, there are enough assembled to cover the land masses of the earth, and the differences in behavior patterns are staggering. But the community, the city, the nation, the world are all different statements composed from the same vocabulary, the one we call man.

Viewed from this perspective, evolution is the translation of one vocabulary into another. We are presently translating our body vocabulary into our cultural vocabulary of words, numbers and musical notes.

Our assumed motivation for translating our selves out of our bodies into thought patterns had been our fear of death. This led us to the awareness that we have always been thought patterns. Through Heisenberg's revelation that atoms are not things but processes we now have evidence that what we call our body is itself a thought pattern composed of thought patterns. We can conjecture that the cosmos composed of stars and planets is composed of thought patterns. Death, we may now understand to be a thought pattern, as fear must also be.

We can define a house, armor, battleships, tanks and airplanes as thought patterns. As are cars, commuter trains, subways, buses, office buildings, factories, airplanes and ships. All the bodies we have been building are thought patterns. However we define them we define them as thought patterns. There is no other way to define them. For example, what is a battleship? A weapon. A weapon is a thought pattern.

What we see is, we have not been doing two different things at once, building bodies for our body to occupy and at the same time seeking an exodus from our body. We have been doing one thing, translating our body vocabulary and our earth vocabulary into our cultural vocabulary of words, numbers and musical notes. Already we have begun the translation of our cosmic vocabulary into words, numbers and musical notes.

This seems to be what we do—we translate one vocabulary into another.

Bridge 10

OUR VERBAL SELF

Werner Heisenberg, winner of the 1932 Nobel Prize in Physics, said: "Atoms are not things but processes."

Sir James Hopwood Jeans, the distinguished English mathematician, physicist and astronomer, said, "Today there is a wide measure of agreement, which on the physical side of science approaches almost unanimity, that the stream of knowledge is heading towards a non-mechanical reality; the universe begins to look more like a great thought than like a great machine."

Sir Arthur Stanley Eddington, the distinguished British astronomer and physicist, said: "The stuff of the world is mind-stuff."

Assuming our cosmos is composed of atoms and

atoms are a process, then our cosmos is a process. According to Sir James Jeans and Sir Arthur Eddington, that process is a thought process.

Webster's New Collegiate Dictionary defines "thought" as: "past of think." "Think" is defined by Webster's New Collegiate Dictionary as: "1: to form or have in mind. 2: to have as an intention...:etc."

If our cosmic process is a thought process; if to think is "to form or have in mind; to have as an intention," it would appear that by cosmos we mean a capacity to form, to have in mind, to intend.

"Form" according to Webster's New Collegiate Dictionary means: "1: to give form or shape to: FASHION 2a: to give a particular shape to: shape or mold into a certain state or after a particular model: ARRANGE (formed the dough into various shapes) (a state formed along the lines of the Roman Republic) b: to arrange themselves in (the women formed a line) c: to model by instruction and discipline (a mind formed by classical education)."

According to this definition of form our cosmic thought process is the capacity to shape, fashion, mold, "into a certain state or after a particular model," arrange, model by instruction and discipline.

"Mind" as defined by Webster's New Collegiate Dictionary is: "1: RECOLLECTION, MEMORY (keep that in mind) (time out of mind) 2a: the element or complex of elements in an individual that feels, perceives, thinks, wills and especially reasons b: the conscious mental events and capabilities in an organism c: the organized conscious and unconscious adaptive mental activity of an organism 3: INTENTION, DESIRE (she changed her mind)."

According to this definition of mind our cosmic thought process is the capacity to remember, feel, perceive, think, will and especially reason, to be conscious and consciously and unconsciously adaptive, to intend and to desire.

"Intend" as defined by Webster's New Collegiate Dictionary means: "la: SIGNIFY, MEAN b: to refer to 2a: to have in mind as a purpose or goal; PLAN b: to design for a specified use or future."

According to this definition of intend our cosmic thought process is a capacity to signify, mean, refer to, be a purpose or goal, a plan, a design for a "specified use or future."

In sum, if our cosmic thought process is a capacity to form, have in mind and intend, then it shapes, fashions, molds, arranges, models by instruction or discipline, remembers, feels, perceives, thinks, wills, reasons. It is conscious, adaptive consciously and unconsciously. And it intends, desires, signifies, means, is a purpose, goal or plan for a specified use or future.

All of this presumably describes us.

When Werner Heisenberg, Sir James Hopwood Jeans and Sir Arthur Stanley Eddington said, "The atom is not a thing but a process"; "the universe begins to look more like a great thought"; "The stuff of the world is mind stuff," they were saying reality is what we think it is. Apparently our thought process is our cosmos. To put it the other way around, our cosmos is our thought process. Still another way of putting it is we are our thought processes.

A thought process is presumably composed of thought processes, composed of thought processes, composed of thought processes. This would indicate we are composed of many different thought processes.

This is different from thinking of ourselves as physical objects apparently unconnected to other unconnected physical objects that composed what we called the physical universe.

When we thought of ourselves as physical objects we thought our words were about a physical, unconnected, objective reality. By that we meant the function of words was to mirror or reflect a visual reality. We thought words were not a reality but symbols for reality. Reality, we thought, was what we saw.

If reality is our thought process, we don't see; we visualize.

When we thought we saw we said, "Seeing is believing." We thought our visualizations were reality and our verbalizations were about our visualizations.

"About" is defined by Webster's New Collegiate Dictionary as: "1: on all sides: AROUND 2a: In rotation b. around the outside 3a: APPROXIMATELY b: ALMOST (about starved) 4: Here and there 5: in the vicinity: NEAR."

Our verbalizations are not on all sides, around, in rotation, around the outside, approximately, almost, here and there, in the vicinity, near, our visualizations.

If we are our reality, our reality is our self.

If we are composed of many different thought processes, if we are our thought processes, we are many different "selfs." Within this context, one of our selves is our verbalization self, another is our visualization self. Our verbalization self is not about our visualization self.

Our verbalization self is different from our visualization self. We cannot ask what our visualization self is. It is a verbal question demanding a verbal answer, asked by our verbal self of our verbal self. For example, the question, "What are our planetary, star, or galactic thoughts?" is an unanswerable question.

We cannot step outside our verbal self to verbalize our understanding of what our verbal self is all about. We either are our verbal self or we are not our verbal self. If we are not our verbal self we are one of our other selves. Some of the categories of our almost infinite number of selves are our feeling self, our musical self, our mathematical self and our visualization self.

We are not simultaneously all of our selves. We are only one self at a time. Whatever self we are, that is the only self we are, the only self that exists.

None of our selves is about another self. None of our selves is about itself. Our feeling self is not about feeling but is feeling. Our musical self is not about music but is music. Our mathematical self is not about mathematics but is mathematics. Our visualization self is not about visualization but is visualization. Our verbal self is not about verbalization but is verbalization.

A thought process is not about a thought process; it is a thought process.

If we are our reality, our reality is not about us but is us. This is a different perspective from thinking of ourself as about something. We thought of our self as about something when we thought of our self as a physical object unconnected to other physical objects.

To think of our self as about something is to think of our self as around, approximately, almost or near some other thing or self. That other thing or self we called God. The way we put it is, God is perfect and we are imperfect.

Webster's New Collegiate Dictionary defines "perfect" as: "1: EXPERT, PROFICIENT (practice makes perfect) 2a: being entirely without fault or defect: FLAWLESS (a perfect crime) b: satisfying all requirements: ACCURATE c: corresponding to an ideal standard or abstract concept (a perfect gentleman) d: faithfully reproducing the original; specifically; LETTER-PERFECT e: legally valid 3a: PURE, TOTAL b: Lacking in no essential detail: COMPLETE, etc."

When we thought of our self as about God the question we asked was, "If God is perfect how could a perfect God make us imperfect?" As the work of his hands it would seem we ought to be perfect as proof of his perfection. Either God is perfect and we are perfect or God is imperfect and we are imperfect.

"God" is defined by Webster's New Collegiate Dictionary as: "1: capitalized: the surpreme or ultimate reality: as a: the Being perfect in power, wisdom and goodness whom men worship as creator and ruler of the universe b: Christian Science: the incorporeal divine Principle ruling over all as eternal Spirit: infinite Mind 2: a being or object believed to have more than natural attributes and powers and to require man's worship; specifically: one controlling a particular aspect or part of reality 3: a person or thing of supreme value 4: a powerful ruler."

When we thought we were about God, we said we were created in his image.

God, we said, could do as he pleased. God represented absolute free will. We, we said, in being about God, are about free will. To be in God's image, we said, meant we could do whatever we chose to do. We could even do what God didn't want us to do.

When we thought of our self as a physical object unconnected to the unconnected physical objects that composed our universe we assumed free will was possible. We thought God's free will created, ruled and controlled all the unconnected physical objects of our unconnected physical object universe, including us. It was this will of God's that did connect all the otherwise unconnected physical objects of this universe. We thought it was our free will that created, ruled and controlled what we called "culture." We thought our free will connected or disconnected all the physical objects, including our selves and other people, within our reach.

The unresolved question within this context was how great or how small was God's control over us? It was generally agreed God controlled our spirit before and after life absolutely. How much he controlled it and our free will during life was debated.

If we are our reality we are not an unconnected physical object in a universe of unconnected physical objects. If our reality is our thought process, a physical object is our thought, a verbal thought.

Within this context the words unconnected physical objects, perfection, imperfection, God, create, rule, control and free will are all part of our verbal self. As are the words atom, form, mind and intend.

To ask "Why?" is to ask what our verbal self is all about. Our verbal self is words. Our verbal self is our verbal self, as our feeling self is our feeling self, as our musical self is our musical self, as our mathematical self is our mathematical self.

We are words. We are feelings. We are music. We are mathematics. Whatever we think, we are.

Our almost infinite number of selves are not about each other, they are our selves.

To think in terms of being about is to think in terms of time. Between what you are and what you are about to be is a gap. That gap is called time.

An implication of the concept that we are not about reality but are reality is there is no gap. There is no time. We are. We never were. We never will be. We are.

Bridge 11

GAMES

Sir James Hopwood Jeans, the distinguished English mathematician, physicist and astronomer, said in his Rede Lectures: "Today there is a wide measure of agreement which on the physical side of science approaches unanimity that the stream of knowledge is heading towards a non-mechanical reality; the universe begins to look more like a great thought than like a machine."

Sir Arthur Stanley Eddington, the distinguished British astronomer and physicist, said: "The stuff of the world is mind stuff."

Apparently the new concept of reality is that it is composed of thoughts, not things. So far as we know these thoughts are our thoughts. Accordingly we must redefine ourself as a capacity to think and the cosmos as composed of our thoughts. From this perspective the cosmos is our way of thinking. This would mean we have star thoughts, planet thoughts, galactic thoughts as well as molecular, atomic and sub- atomic thoughts.

Apparently reality is what we think.

What we have called "material" is our three-dimensional way of thinking. We may think in terms of three dimensions but we cannot step outside our thoughts and view them objectively. We are either in one thought or in another. To put it another way, thoughts do have an inside but do not have an outside. The fact we once thought they did indicates we thought differently then than we do now.

According to our three-dimensional, "material" way of thinking "reality" is an objective thing. We thought the stars, the planets, the earth and all that was on it including our bodies were things outside our thoughts. Everything outside our thoughts we thought composed one objective reality. We thought this one objective reality could be measured by one measuring stick. Our motivation for measurement was our hope of being able to predict absolutely what would happen. This is cause and effect thinking: it means, if this happens, that has to happen.

When Werner Heisenberg discovered the atom was not a thing but a process he conceptualized what came to be called the principle of uncertainty. He was given the Nobel Prize in Physics in 1932 for this discovery.

Since no one has ever seen an atom, the atom has always been a concept. Heisenberg changed our concept of the atom as a thing to the concept of a process. All that has changed is our thinking. The atom is still a thought, only a different thought. An implication of this different thought is that reality is what we think.

Since the atom, the assumed building block of our "thing reality," is not a thing, there is no basis for a belief in cause and effect thinking. This means we can no longer hope to predict absolutely what will happen.

While we think "things" are predictable, we know our thoughts are unpredictable. What Heisenberg's principle of uncertainty appears to indicate is, our reality is as unpredictable as our thoughts. The atomic process Heisenberg discovered is one of our thought processes.

The word "logic" is a key word in a cosmos composed of thought processes. Each thought process has its own system of logic. Logic is structured in terms of laws. Games are structured in terms of laws or rules. Games are composed of specific systems of logic. From this perspective stars, planets, galaxies as well as molecules, atoms and neutrinos may be viewed as games. Each game has different laws representing different systems of logic.

According to this perspective, "reality" is composed of games which are composed of games which are composed of games. Each game describes a different way of thinking.

In general this indicates the complexity of our thought. In particular it indicates the question "What does it all mean?" is an answerless question. To answer this question would necessitate defining all our games in terms of one game. Our game baseball cannot be defined in terms of our game Mars.

What we seem to do is synthesize two or more games into a new game. We call this process of synthesis evolution. Having done this we have not found one meaning for two different games or systems of logic. We have evolved a new meaning relating exclusively to the context of that new game.

Each game defines its own meaning according to its own system of logic. The meaning of one game is meaningless when applied to another game.

The difference between games is apparent in their different systems of logic. In general, the system of logic of planets is different from that of novels, orange trees, violins or people. In particular the system of logic of the planet Mars is different from that of the planet earth. In particular each novel represents a different system of logic, as does each orange tree, violin or person.

According to our three-dimensional way of thinking we thought all games or systems of logic, except our human thought and recreational games, were things. We defined the difference between things in terms of measurements.

Assuming our universe to be one thing, it seemed reasonable to assume we could measure it and its parts by one ruler. Our ruler measured length, width, height, weight and time, or how long a thing lasted. If, as in the case of the novel, we couldn't meaningfully apply these measurements we assumed the novel was less real than, say, an orange tree. In effect we felt if we couldn't measure it, it didn't exist. By this we meant it was illogical according to the logic of our three-dimensional system of logic or game.

If what we call reality is a game or system of logic, it would follow there are as many realities or systems of logic as there are games.

Within this context no reality is more nor less real than any other reality. Putting it another way, no system of logic is more nor less logical than any other. Perhaps a clearer way of putting it is, each reality or system of logic has a unique pattern. No pattern is more nor less of a pattern than any other pattern.

If our reality is composed of our thoughts, each pattern represents a different way of thinking.

To think in terms of three dimensions is to measure. This indicates a concept we call "direction" which indicates such concepts as "here" and "there." Our motivation for measuring is to prepare for the future. This indicates a concept we call the "future" which implies such concepts as "past" and "present."

To think in terms of direction and future is to think in terms of values. The questions "Where do I come from?" "Where am I going?" and "What will become of me?" are questions of value. Our philosophies and religions are value systems.

The logic of our three-dimensional way of thinking is the logic of values.

"Value" is defined by Webster's New Collegiate Dictionary as:"1: a fair return or equivalent in goods, services, or money or something exchanged. 2: the monetary worth of something: marketable price 3: relative worth, utility, or importance: degree of excellence (had nothing of value to say) 4: a numerical quantity assigned or computed b: the amount or extent of a specified measurement of space, time, or quantity c: precise signification (value of a word)."

According to Webster's New Dictionary a value system is a system of measurement.

Our three dimension game or system of logic appears to be one of measurement or values.

The question then is, "Why did we evolve that particular game?" Why did we conceive of a value way of thinking?

"Why" is a question that can only be asked and answered within the game of values. "Why" is a teological question.

Webster's New Collegiate Dictionary defines "teology" as: "1a: the study of evidences of design in nature b: a doctrine (as in vitalism) that ends are immanent in nature c: a doctrine explaining phenomena by final causes 2: the fact of character attributed to nature or natural process of being directed toward an end or shaped by a purpose."

In other words value thinking is thinking in terms of ultimate ends. The assumption of value thinking is that our reality has an end. If reality is what we think, to think in terms of values is to assume our capacity to think is finite.

Webster's New Collegiate Dictionary defines "finite" as:"1a: having definite or definable limits (finite number of possibilities) (a finite community) b: having a limited nature or existence (finite beings) 2: completely determinable in theory or in fact by counting, measurement or thought: neither infinite nor infinitesimal (a finite distance) (the finite velocity of light)."

To ask "why" is to assume our reality is finite, measurable, completely determinable. It is to assume our reality is a value system. Such assumptions logically evolved from our three-dimensional way of thinking. The name of this game or system of logic is "materialism." According to this game reality is an objective thing composed of things.

According to Werner Heisenberg our reality is not composed of things. He said, "The atom is not a thing but a process." According to both Sir James Hopwood Jeans and Sir Arthur Stanley Eddington our reality is composed of thoughts.

If reality is composed of thoughts, it is not material, measurable, nor three-dimensional. To think it is, is to think in terms of a value system of logic. If reality is composed of thoughts it is not determinable for we don't know what we will think. The thoughts we do think are definable in terms of its unique system of logic. To think of our thoughts as games is to understand we do not choose our reality. We do appear to choose how well we intend to play the games, the reality, we do think.

Bridge 12

GOD

The word "God" has been a symbol for many different ideas. We've meant control (who's in charge), fate, reward as well as punishment, judgment, the source, the ultimate, the best, caring, the future, but in particular we have meant our creator—the responsible parent being. "God" as creator has been our way of saying our lives make sense, have meaning.

This tells us a lot about ourselves. Clearly, we care very much that our experiences make sense, that even if we don't understand life, it means something. What we believe is, if life is meaningful we are—if life is not meaningful, we are not meaningful. We apparently associate our sense of living, being, with a sense of meaning. Where we are unaware of any meaning we say the situation is chaotic. The essence of terror is the acceptance of the possibility of meaninglessness, for within this acceptance rests the possibility of our no longer being. With us being and meaning are interchangeable. To say I am is to say I mean something. To say I am not is to say I mean nothing.

But who can believe that? We can fear meaninglessness but we cannot accept it. We'd rather die than accept it. To put it another way, we are prepared to kill anyone or thing we suspect capable of depriving us of our meaning. Such a reaction indicates we are responding to an accepted threat to our being. Not our lives. We don't take our lives that seriously. If we did we would never dare to do anything for practically everything we do has the potential to kill us. No, it is our meaning, our being, that concerns us, not our lives. Why else would we fight wars, or duels as we did in the past? Why would we fight at all? To say we do it for money is sheer nonsense. We value money because it seems to give us meaning or, as it is sometimes expressed, money seems to fulfill our desire to be "somebody." Wars fought for religious or political reasons are motivated by the fear some nation or group of nations can take our meaning, our very being away from us. Clearly the way to end all wars, indeed all strife, is to take our meaning or being out of our hands, to realize that it is not ours to give or take. In conceiving of God-the-creator we thought we were doing this.

The idea of God-the-creator was a good idea, a step in the right direction, for it provided two assurances. Having a creator established that we, indeed everything, was meaningful and it established that meaning or being was not ours to give or take away. Creating a creator in our own image didn't work because many nations conceived of many God-creators. Our interest in solving the resulting conflicts is manifested in today's ecumenical movement which in essence is an attempt to synthesize all concepts of God into one God-creator.

Another solution to the conflict of many gods has been the elimination of the concept of God, the giver of meaning or being, entirely. Many people consciously or unconsciously have replaced the concept of God-the-creator with the concept of "the people," "society" as the givers of meaning. In doing so we equated meaning with popularity. The more popular we are, the more meaningful and the more real we are. To understand this in the free world is to understand that democracy somehow came to signify we-the-people are the ultimate givers and takers of the individual's meaning or being. In totalitarian states, the concept of God-the-creator is replaced by the state as the giver of being or meaning. But the state claims that all that it does it does in the name of the people. So both democracies and totalitarian states accept the concept that it is the people who are the givers of meaning. For capitalistic countries the "marketplace" has been designated as the arena all citizens must enter to fight to be granted the gift of meaning. In totalitarian states the arena in which the individual fights for the gift of meaning is the bureaucracy.

First we put our meaning or being beyond our reach in the hands of God-the-creator and when that didn't seem to work we took it back and gave it to the people and the state. But in so far as our meaning or being could still be given or taken from us nothing much has changed.

The problem of whether our meaning or being can be given or taken from us centers on the question of creation.

What would happen if we said we weren't created?

According to Webster's New Collegiate Dictionary, to create is "to bring into existence." If we said we were not created we would then be accepting ourselves as having always been. We would be saying we are timeless. We would be saying we are not finite. We would be saying we are immeasurable, that is infinite. Since all that can be given or taken is finite, if we are infinite we cannot be given or taken. Then neither could we be understood, which would mean that to be infinite would be to be meaningless to finite perception. From this perspective to accept ourselves as meaningful is to accept ourselves as finite, that is something that can be given and taken away.

So by the word "creation" we have meant the process of finitization. Therefore any story of genesis of the world is a story of finitization. And in fact that is how such stories read, saying that in the beginning there was nothing out of which was created something. But what was the nothing from which something was created? If we are infinite it was us, and the something we created was thought. What else could it be?

According to Plato the concept of creator is intended to account for the "intelligibility" of the universe. Which is another way of saying the universe is, in fact, our understanding of our self. What we see then is that to understand ourself is to "create." Another way of putting it is that in the beginning we thought and what we call the universe is composed of our thoughts. So by "the beginning of the universe" we mean the beginning of our thoughts. It isn't that we began then: what began then is our thoughts.

The fact is the universe is what we think it is. We only have to look back at history to know that. What we call history is the story of our thoughts we call the universe. The differences between our ancient Egyptian universe and our Chinese universe, for example, and our contemporary universe are striking. It isn't that we were more ignorant then than now: it's that we think different thoughts now than we did then. What we call evolution is the history of our ever-changing thoughts.

It presently appears that initially our thoughts were non-visual, composing the component parts of the thought process we call the atom. At some point we synthesized from vast numbers of these atomic thoughts our visualization-thought process. We call this light. What we call seeing is "visualizing." Our first visualization may well have been of a vast glowing hydrogen cloud billions of miles long and billions of miles wide from which condensed our first star-thought, or perhaps our first galactic-thought. We are trying to remember how we did it, which is what astronomy is all about. From this point of view all stars are memories of the past and it is the study of these memories we call astronomy.

At a later point we synthesized from our visualizations a way of thinking we called "man." Through this man-thought process we are evolving a new cosmos composed of words, numbers, musical notes, what might be called a cultural cosmos for lack of a better name. While we do still visualize the indications are we are evolving out of that particular thought process.

We are, through what we call science, trying to remember how we did it all. Or so we imagine. In fact, what appears to be happening is we are in the process of synthesizing all our previous thoughts into a new thought. Just as we did when we synthesized our process of visualization which we call light from all our thoughts that preceded it.

The word "God" is our finite symbol for our infinity.